YOUR KNOWLEDGE HAS VALUE

- We will publish your bachelor's and master's thesis, essays and papers

- Your own eBook and book - sold worldwide in all relevant shops

- Earn money with each sale

Upload your text at www.GRIN.com and publish for free

Ataliba Miguel

Heat exchanger failure investigation report

GRIN Verlag

The Robert Gordon University

HEAT EXCHANGER INVESTIGATION FAILURE REPORT

Materials & Corrosion Science Technical Report

Ataliba Miguel
8/1/2013
Word Count 1937

ABSTRACT

A high pressure gas cooler located in an offshore platform have been operating for more than 10 years. Throughout that period the gas cooler have been subjected to several tube failures, the failures have caused gas leak from the tube side. Several materials upgrades have been used to contain the tube failure. The last choice was to use a more corrosion resistant material Hastelloy C22. Crevice corrosion has been reported as the primary failure mechanism. The tube and tube plate joined surfaces have been exposed to high temperature which is relatively higher than the critical crevice temperature of Hastelloy C22. There was a poor heat transfer between the shell side fluid and tube side due to a small heat transfer area and low fluid velocity in the affected zone. Stress corrosion and fatigue corrosion accounted for the secondary failure mechanism which ultimately caused a crack in the tubes. Other possible reason identified to cause crevice corrosion was the mechanical rolling expansion technique. Oftentimes it creates rear crevices on the tubes with enough geometry to develop crevice chemistry. Mitigation methods such as hydraulic expansion technique, heat treatment for residual stresses and baffle design enhancements have been proposed in this report.

Key words: shell and tube heat exchanger failure; Hastelloy C22 material; tube failure; localized corrosion; crevice corrosion; internal pitting corrosion

<u>List of Contents</u>

LIST OF IMAGES

List of Tables

ABBREVIATIONS
API   AMERICAN PETROLEUM INSTITUTE
CCT   CRITICAL CREVICE TEMPERATURE
CH_4 METHANE
CPT   CRITICAL PITTING TEMPERATURE
HCl   HYDROCHLORIC ACID
HP   HIGH PRESSURE
NaCl   SODIUM CHLORIDE
SCC   STRESS CORROSION CRACKING

1. Introduction

A HP gas (CH$_4$) cooler heat exchanger of the shell and tube type located in an offshore platform suffered a tube leak failure after six months short service period since it was last repaired from the previous failures. The heat exchanger have experienced from the past ten years several materials upgrade to contain the failures. A defect assessment of the tube bundle was carried out by eddy current, several sketches and photographs were provided to aid describing the nature, reasons and the general circumstances of the failure.

1.1.   Methodology

The methodology adopted to carry out this failure investigation was based on the available information supplied by the client. It is worth to point out that there was a lack of some useful information needed to aid carrying out this failure investigation. Therefore, all the information not readily available by the client was hypothesized to fit the evidence seen from the photographs.

2. Failure Investigation

A detailed description of this failure investigation is briefly explained in the following subsections.

2.1.   Heat Exchanger material description

The material used for the tube bundle was a highly corrosion resistant nickel-chromium-molybdenum alloy C22. Such austenitic alloys rely on the stability of a thin chromium oxide film for protection against corrosion (Mon, Gordon and Rebak, 2005). The chemical composition of C22 alloy is listed in table 1. The material used for shell and tube plate was a low alloy steel.

Table 1 – Hastelloy C22 composition

Nominal composition, wt%									
Ni	Co	Cr	Mo	W	Fe	Si	Mn	C	V
56	2.5	22	13	3	3	0.08	0.5	0.01	0.35

2.2.   Service condition

The table 2 below refers to the design and operating temperature of the
heat exchanger.

Table 2 – Stream temperatures

Seawater		
	Design	Operating
Cooling water in	18 C	16.7 C
Cooling water out	29.4 C	42.4 C
HP Gas		
	Design	Operating
HP Gas in	169 C	160 C
HP Gas out	49 C	40 C

2.3   Localized corrosion

Seen from the photographs the onset of tube failure was due to localized
corrosion. The localized corrosion occurred at location of the joined metal
to metal surface between the tube and tube plate exposed to high
temperature. At that location there was a possible dead zone with limited
flow of coolant fluid, producing a poor heat transfer, between the shell
side fluid and the tube side. Within the metal to metal surface there was
a gap with restricted geometry which supported the development of
crevice chemistry (Shan and Payer 2008).

Figure 1 – Localized crevice corrosion row 19 – tube 1 r.h.s

2.4   Effect of rolling expansion

The process of de-stubbing and re-tubing was an operation performed several times during the service life of the heat exchanger.

The mechanical roller expansion method induces a radial force inside the tubes. This increasing force moves the tube material outwards until it contacts the internal diameter of the tube plate hole and continues until the tube plate material is just below its yield point (Bloodworth).

The process of re-tubing into the tube plate holes, may result in deterioration in the strength of tube to tube-plate joint having large initial clearance (see figure 3) as well as in the strain hardening and thinning of the expanded tubes and their surrounded ligaments (Merah et al., 2012).

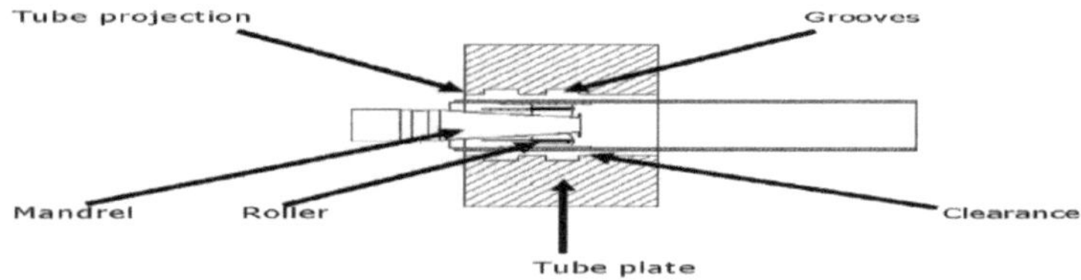

Figure 2 – Rolling expansion

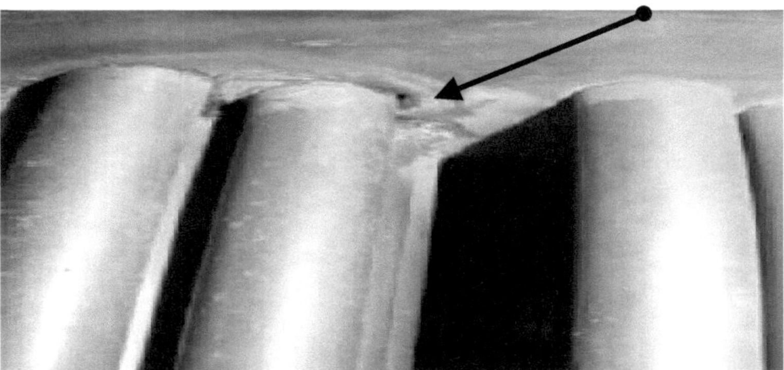

Figure 3 – Deterioration of the strength in the tube to tube-plate joint.

2.5   Stress corrosion

Stress corrosion cracking is the combined influence of tensile stresses when subjected into a corrosive environment. The impact of stress corrosion induces the material to a crack. The required tensile stresses are in the form of residual stresses. The residual stresses occur due to inelastic deformations and heat treatment. When rolling expanding the tubes, the differential displacement of the inside diameter surface of the tubes in the transition zone from the fully expanded to unexpanded zones creates both a tensile and compressive residual stresses (Merah et al., 2012)

2.6 Corrosion fatigue

Taking into consideration that heat exchangers are subjected to thermal expansions in the tube bundle, it is thought that fatigue corrosion also played a role on the crack initiation, as a result of repeated cycling of thermal stresses below the material static yield strength. The crack as shown in figure 4 propagated in the region where the strain is most severe.

Figure 4 – crack initiation as a result of combined stress and fatigue corrosion. Row 18 tube 2 r.h.s.

2.7 Effects of temperature

The gas inlet temperature at the location of failure was considerably high. Hastelloy C22 material have sufficient alloying to be resistant to crevice and pitting attack up to a certain temperature. Laboratory test have been performed to evaluate the critical crevice and pitting temperature using a

NaCl-HCl solution. The values obtained from the laboratory test were respectively 102 °C critical crevice temperature and 150 °C critical pitting temperature (Haynes Intl). Information obtained from laboratory data suggest that HASTELLOY C-22 is prone to corrode minimally at 160° C, provided the corrodant is only seawater with no additional acidification. Email confirmation was received (D Metzler 2013, pers. comm. 15 July).

3. Discussion

The technical reasons which led to tube leak failure were analysed and thoroughly explained in the next subsections.

3.1.   Causes of crevice corrosion

Crevice corrosion is highly susceptible to occur between the tube and tube plate metal to metal joined surfaces. One of the root cause deduced during this investigation was due to the rolled expansion method adopted to assemble the C22 tubes in the grooves recesses of the tube plate. Figure 4 highlights an incipient crevice corrosion which is thought to be linked with the rolling expansion tube. The rolling expanded method in the grooves recesses pushes out the tubes, which oftentimes creates a crevice in the tubes.

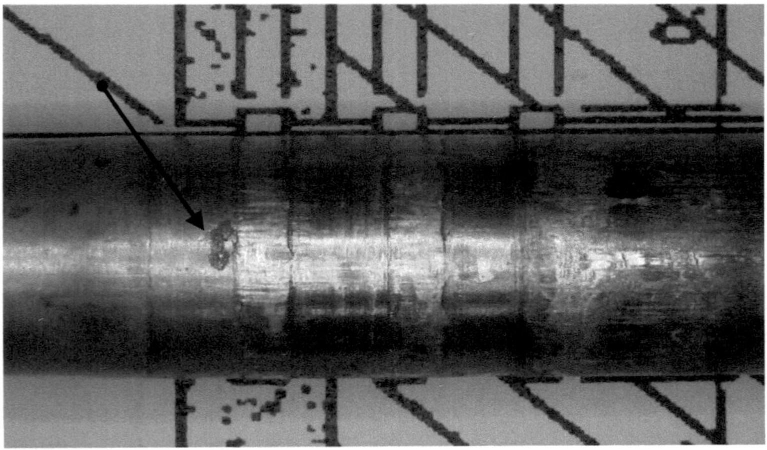

Figure 5 – incipient crevice corrosion. Row 6 – tube 4 l.h.s.

3.2. Crevice corrosion mechanism

Crevice corrosion occurs within the occluded regions of a metal to metal surface (tubes to tube plate), and requires the presence of a continuous liquid film connecting the occluded region to the external bulk material. Within this region there is a volume of stagnant liquid inside a crevice with enough geometry to allow liquid entry but narrow enough to maintain a stagnant layer. The initiation of crevice corrosion involves the local depletion of oxygen in the crevice, a spatial separation of the anode from the cathode, a metal dissolution and the hydrolysis of dissolved metal ions and the accompanying acidification of the creviced region (Mon, Gordon and Rebak, 2005). Within the crevice the pH is low whereas there is a high concentration of chloride ions that neutralizes the charge imbalance promoting a local breakdown of passive films which leads to an increased metal dissolution rate.

Figure 6 – Localized crevice corrosion. Row 15 – tube 12 r.h.s.

3.3. Stream temperature effects

Table 2 shows the process stream temperatures. The temperature approach is the difference between the hot side and cold side fluid temperatures at any point within a heat exchanger (KLM Group, 2010). Thus, the values given on the table clearly shows that there is a greater temperature difference between the cold side outlet fluid stream and, the hot side inlet fluid stream. As a general guide the greater temperature difference between both streams should be 20 C and the least should be 5 to 7 C (Sinnott and Towler, 2009). As seen the HP inlet gas is 160 C and the outlet cooling water is 42.4 C, the temperature difference is 117.6 C which is still above the CCT for Hastelloy C22 within that region.

3.4.   Causes of pitting

When the metal is exposed to an oxidizing gas at a considerable high temperature, corrosion occurs by direct reaction with the gas. Figure 7 shows some deposits of iron oxide inside the tubes. The rate of attack tends to increase substantially at increased temperature. The surface film thickens as a result of reaction scale/gas or metal/gas interface due to cation or anion transport through a solid electrolyte (Dokumaci, 2006). Once the passive film inside the tube is breached, an electrochemical cell becomes active. Iron goes into solution in the more anodic bottom of the pit, diffuses toward the top, and oxidizes to iron oxide (Schiroky et al., 2013). As such, pitting occurs causing a rapid penetration in the metal substrate. Pitting it is characterized as a corrosive attack in a localized region caused by the potential difference between the anodic area inside the pit and, the surrounding cathodic area (Jacobson, 2010).

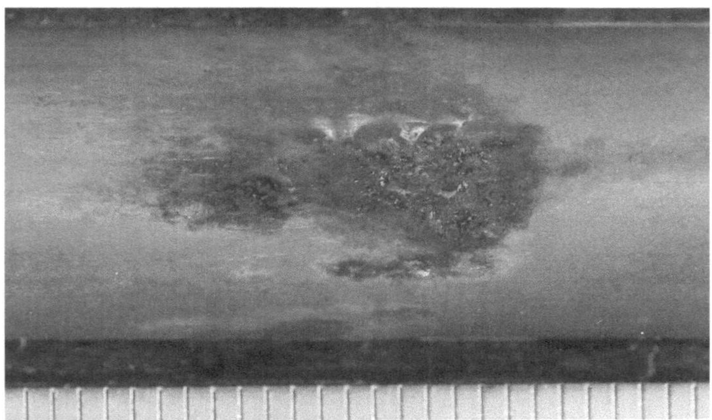

Figure 7 – Iron oxide deposits. Row 8 tube 3 r.h.s.

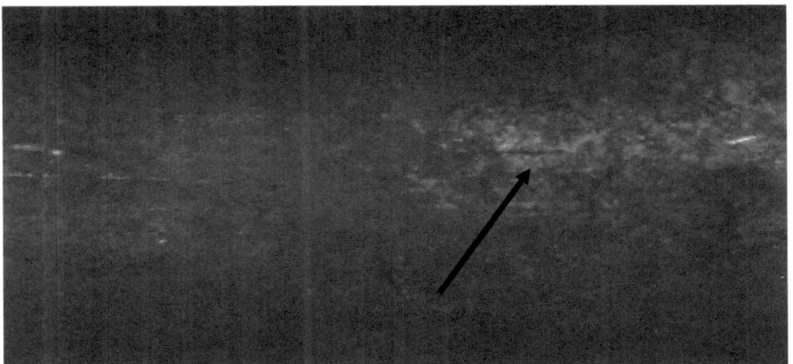

Figure 8 – Internal pitting underneath iron oxide deposit. Row 8 tube 3 r.h.s.

3.5.   Crack initiation

In the region where failure occurred both SCC and fatigue corrosion were the competing mechanisms. The tubes have being exposed to both cyclic loading and static tensile loading in a corrosive environment (Warke, 2002). Vibration problems are commonest in single segmental baffle type shell heat exchanger. SCC in heat exchangers is encouraged by open crevices and secondary side residual stress due to cold work on the outer surface of the critical transition between the expanded and unexpanded tube (McGregor, 1999).

Tubes subjected to tensile stresses and fatigue loads, have their protective surface film ruptured as a result of a combined event. The site of crack initiation causes the passive layer on the surface of the metal to be locally thinned, allowing pitting or grain boundary corrosion to start. Stress raisers results frequently from spalling or localized corrosion such as pitting and crevice corrosion (Warke, 2002).

Figure 9 – Crack on the tube. Row 17 tube 2 r.h.s.

4.   Conclusions

Failure of the tubes occurred as a result of several corrosion mechanisms.
Crevice corrosion was the primary mode of tube failure; the reason is
linked to mechanical rolling expansion technique, and the temperature
difference between hot side inlet stream and cold side outlet stream
which is considerably higher as compared to CCT of Hastelloy C22. SCC
and corrosion fatigue accounted for the secondary mechanism of tube
failure which promoted the initiation of a crack.

5.   Recommendations to mitigate failure

The following subsections consider some of steps to be adopted to
mitigate the major sources of corrosion.

5.1   Tube assembly

Connection of the tubes and tube plate represents the weakest part in
the heat exchanger structure and is the main source for failure.
To avoid crevice corrosion, tubes should be contact expanded into the
tubesheet for a length of the tubesheet thickness minus 3 mm (API 660).
An effective tightness of the contact surface between the tubes to
tubesheet is achieved by hydraulic expansion. Hydraulic expansion
ensures that residual contact pressure is high enough to keep tightness

of the joint (Qian et al., 2006). A good joint tightness adherence eliminates stagnation areas of the fluid flow.

5.2   Residual stresses

Twisted U tube bend possesses substantial cold work and residual stresses before heat treatment. Therefore before any heat treatment is given, it is recommended to relieve the stresses by using a solution annealing treatment. The resistance solution-annealing treatment seems particularly beneficial, since it imparts a residual stress pattern of either compressive or very small tensile stresses (Mahajanam, McIntyre and Hovey, 2009).

5.3   Temperatures and baffle design

Taking into consideration the temperature approach is relatively higher than the CCT for Hastelloy C22, it is recommended to design a heat exchanger with a large area for heat transfer, or a high fluid velocities to increase the overall heat transfer for the given duty design. Single segmental baffles present some deficiencies such as potential for dead zones and excessive tube vibration. Therefore it is recommended to use baffle enhancements such as helical baffles to attempt alleviate the problems (Lunsford, 1998).

6.　References

API 660. (2007) "Shell and tube heat exchanger". *ANSI/API Standard 660.* [Online] Available at:
http://subscriptions.techstreet.com/searches/607548
Accessed on 19th July 2013.

Bloodworth. M. (unknown year) "Tube expansion issues and methods". *Haskel International Inc.* [Online]. Available at: http://www.haskel-europe.com/Files/Haskel-UK/Global/US-en/Tube_Expansion_Issues_and_Methods.pdf
Accessed at 10th July 2013

Dokumaci. E. (2006) "Investigation of high and low temperature corrosion behaviour of nickel-base alloys". Published Msc thesis. [Online]. Available at: http://www.belgeler.com/blg/10r0/investigation-of-high-and-low-temperature-corrosion-behaviour-of-nickel-base-alloys-nikel-esasli-alasimlarin-yuksek-ve-dusuk-sicaklik-davranislarinin-incelenmesi
Accessed at 26th July 2013.

Haynes Intl. (2002) "HASTELLOY® C-22® alloy". *HAYNES Corrosion resistant alloys.* [Online] Available at:
http://www.haynesintl.com/pdf/h2019.pdf
Accessed at 1st July 2013.

Jacobson. A. G., (2010) "Corrosion – A natural but controllable process". *NACE.* [Online]. Available at:
http://events.nace.org/library/articles/corrosion101.asp
Accessed at 18th July 2013.

KLM Technology Group (2010) "Heat exchanger selection and sizing" *Practical Engineering Guidelines for Processing Plant Solutions.* [Online]. Available at:
http://kolmetz.com/pdf/EDG/ENGINEERING%20DESIGN%20GUIDELINE-%20HX%20Rev%203.pdf
Accessed on 30th July 2013.

Lunsford. M. K. (1998) "Increasing heat exchanger performance" *Hydrocarbon Engineering.* [Online]. Available at:
http://www.bre.com/portals/0/technicalarticles/Increasing%20Heat%20Exchanger%20Performance.pdf
Accessed 30th July 2013.

Mahajanam. S.P.V., MczIntyre. R. D., and Hovey. K.L., (2009) "Residual stress control for prevention of environmental cracking of stainless stells". *NACE International.* Paper No 09299. [Online]. Available at
www.onepetro.org
Accessed on 23rd July 2013.

McGregor. R., (1999) "Nuclear steam generator tube to tubesheet joint optmization". *7th International Conference on Nuclear Engineering.* ICONE-7478, pp. 19-23.
http://www.jsme.or.jp/monograph/pes/1999/ICONE7/PAPERS/TRACK02/FP7478.PDF

Merah. N. et al., (2012). "3D finite element analysis of roller expanded heat exchanger tubes in over enlarged tubesheet holes". *Applied Petrochemical Research.* pp.45-52. [Online] Available at:
www.springerlink.com
Accessed on 19th July 2013.

Metzler, D 2013, email 15 July, dmetzler@haynesintl.com

Mon. G. K., Gordon. M. G. and Rebak. B. R. (2005). "Stifling of crevice corrosion in alloy 22". In: T.R. Allen, P.J. King and L. Nelson eds. *The Minerals, Metals and Materials Society.* [Online]. Available at:
http://iweb.tms.org/NM/environdegXII/1431.pdf
Accessed on 1st July 2013.

Namduri. H., (2007). "Formation and quantification of corrosion deposits in the power industry". Published PhD thesis. University of North Texas. [Online] Available at:
http://digital.library.unt.edu/ark:/67531/metadc3635/m2/1/high_res_d/dissertation.pdf
Accessed on 15th July 2013.

Qian. C., et al., (2006). "Reliability study of hydraulically expanded tue-to-tubesheet joint". [Online] Available at: http://www.paper.edu.cn
Accessed on 7th July 2013.

Revie. W.R. and Uhlig. H.H. (2008). " Corrosion and Corrosion Control – An introduction to corrosion science and engineering". 4ed. New Jersey John Wiley & Sons.

Schiroky. G., et al., (2013) "Pitting and crevice corrosion of offshore stainless steel". *Offshore Magazine.* [Online]. Available at:
http://www.offshore-mag.com/articles/print/volume-73/issue-5/productions-operations/pitting-and-crevice-corrosion-of-offshore-stainless-steel-tubing.html
Accessed on 18th July 2013

Shan. X. and Payer. H.J. (2008). "Effect of crevice former on the evolution of crevice damage". *NACE International.* Paper No 08575. [Online]. Available at: www.onepetro.org
Accessed on 3rd July 2013.

Sharpe, R. 2013, email 07 July, Rob.Sharpe@kbr.com

Sinnott. R. and Towler. G. (2009) "Chemical Engineering Design". 5ed. Oxford. Butterworth – Heinemann.

Warke. R. W., (2002) "Stress-Corrosion Cracking". In: W. T. Becker and R.J. Shipley eds. ASM Handbook. Vol 11. pp.823-860 [Online] Available at: www.knovel.com

Zhang. J.F. He. Y.L. and Tao. W.Q. (2010) "A design and rating method for shell and tube heat exchangers with helical baffles". *Journal of Heat Transfer*. ASME. Vol 132.

7. Appendices

<u>Heat exchanger sketch</u>

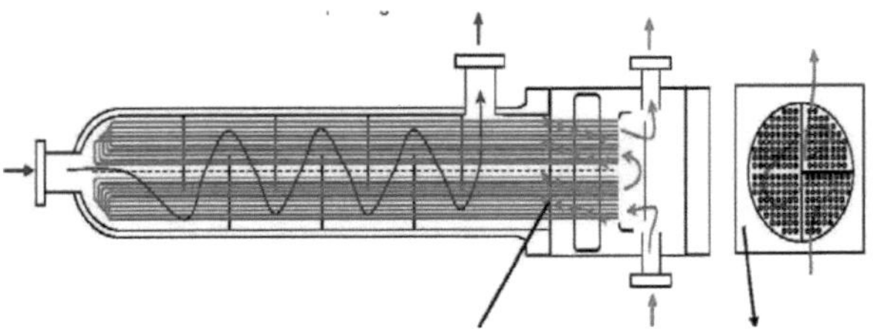

Defect assessment of tube bundle by Eddy current

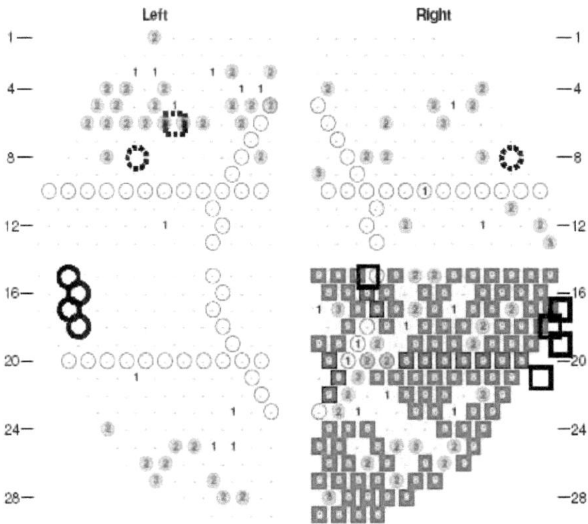

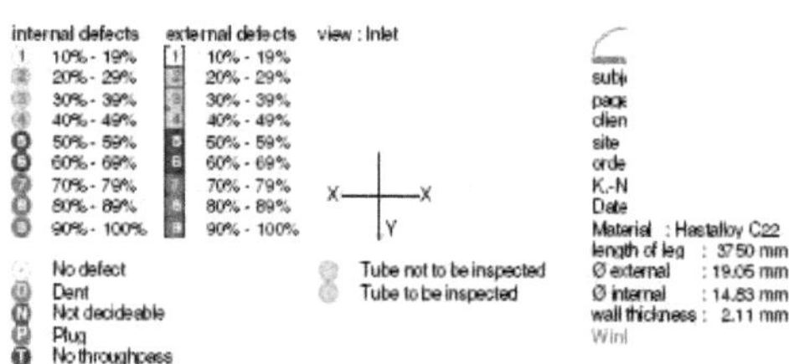

left is referred in text as l.h.s. (left hand side)
right is referred in text as r.h.s. (right hand side)

samples for chemical analysis
samples for crevice corrosion examination
samples for metallographic examination of the internal defect
sample for incipient crevice corrosion examination

Bibliografische Information der Deutschen Nationalbibliothek:

Die Deutsche Bibliothek verzeichnet diese Publikation in der Deutschen National-
bibliografie; detaillierte bibliografische Daten sind im Internet über http://dnb.d-
nb.de/ abrufbar.

Dieses Werk sowie alle darin enthaltenen einzelnen Beiträge und Abbildungen
sind urheberrechtlich geschützt. Jede Verwertung, die nicht ausdrücklich vom
Urheberrechtsschutz zugelassen ist, bedarf der vorherigen Zustimmung des Verla-
ges. Das gilt insbesondere für Vervielfältigungen, Bearbeitungen, Übersetzungen,
Mikroverfilmungen, Auswertungen durch Datenbanken und für die Einspeicherung
und Verarbeitung in elektronische Systeme. Alle Rechte, auch die des auszugsweisen
Nachdrucks, der fotomechanischen Wiedergabe (einschließlich Mikrokopie) sowie
der Auswertung durch Datenbanken oder ähnliche Einrichtungen, vorbehalten.

Imprint:

Copyright © 2013 GRIN Verlag GmbH
Druck und Bindung: Books on Demand GmbH, Norderstedt Germany
ISBN: 978-3-656-61175-2

This book at GRIN:

http://www.grin.com/en/e-book/269681/heat-exchanger-failure-investigation-report

GRIN - Your knowledge has value

Der GRIN Verlag publiziert seit 1998 wissenschaftliche Arbeiten von Studenten, Hochschullehrern und anderen Akademikern als eBook und gedrucktes Buch. Die Verlagswebsite www.grin.com ist die ideale Plattform zur Veröffentlichung von Hausarbeiten, Abschlussarbeiten, wissenschaftlichen Aufsätzen, Dissertationen und Fachbüchern.

Visit us on the internet:

http://www.grin.com/

http://www.facebook.com/grincom

http://www.twitter.com/grin_com